FLORA OF TROPICAL EAST AFRICA

TYPHACEAE

D. M. NAPPER

Rhizomatous herbs with erect unbranched stems. Leaves broadly linear to narrowly elliptic with a long open sheathing base. Inflorescence spike-like, protandrous with contiguous or remote superposed spikes, the upper ♂, the lower ♀, each with a usually caducous bract at the base, sometimes interrupted by similar bracts. Male flowers reduced to 2–5 stamens; flowers interspersed with linear to laciniate bracteoles. Female flowers sterile and fertile together, with or without a subtending bracteole, 1–many on crowded short lateral stumps or " pedicels "; fertile flowers with numerous long perigonous hairs and a long-stalked unilocular 1-ovulate ovary bearing a linear style and linear or lanceolate stigma; gynoecium of the sterile flowers modified to a stout clavate carpodium with a terminal reduced style. Fruit a 1-seeded follicle dehiscing longitudinally.

A monogeneric family occurring in all parts of the world.

TYPHA

L., Sp. Pl.: 971 (1753) & Gen. Pl., ed. 5: 418 (1754); Schnizlein, Typhaceae (1845); Rohrbach in Verh. Bot. Ver. Brandenb. 11: 67–104 (1870)

Stout glabrous marsh herbs with sympodial branched rhizomes covered by distichous cataphylls and having terminal erect shoots. Lower leaves transitional, short-bladed; upper leaves with a long sheath closely enveloping the stem, often auricled at the junction with the blade, and a linear obtuse or acute blade gradually widening from a narrow subpetiolar base, semi-cylindric to lenticular in section, sometimes keeled beneath. Male inflorescence apical, dense; flowers reduced to a pedicellate group of 2–4(–7) stamens with the filaments fused below; anthers dithecous, oblong, with the connective produced above; bracteoles filiform, spathulate or spathulate-laciniate; pollen grains shed singly or in tetrads. Female inflorescence a single spike beneath the ♂, remote from it or contiguous, rarely with 1 or 2 additional ♀ spikes remote from each other beneath the first; flowers with the perigonous hairs and bracteoles of similar length, the styles conspicuously longer; bracteoles filiform, broadening into an expanded ovate or spathulate lamina; carpodia about as long as the bracteoles. Follicle narrowly ellipsoid; seed fusiform.

A small genus of 10–12 aquatic species with wide distributions. Of the four species recorded from tropical Africa, *T. elephantina* Roxb., with a scindo-saharian distribution, is the only one absent from East Africa.

Spikes remote, of equal length or the ♂ the longer*;
 bracteoles present in ♀ spike; stigma linear;
 bracteoles of ♂ spike red-brown, forked; tip of

* *T. capensis* sometimes has remote spikes. Such specimens may be separated from *T. domingensis* by the leaf-sheath, the stigma and the stamen-tip.

stamen dark, rounded; leaf-sheath with sloping
shoulders, purple-spotted within and on the base
of the blade .   .   .   .   1. *T. domingensis*
Spikes contiguous, the ♂ usually much shorter than the
♀; bracteoles absent from ♀ spike or rare; stigma
lanceolate; leaf-sheath with auriculate or rounded
shoulders, often purple-spotted but not on the blade:
Female spike reddish-chestnut at maturity, 1·4–2 cm.
in diameter; bracteoles of ♂ spike red-brown,
flattened above and variously forked; tip of
stamen dark, obtusely triangular; pollen grains
solitary   .   .   .   .   2. *T. capensis*
Female spike yellow-green becoming dark sepia-
brown, 2·25–4 cm. in diameter; bracteoles of ♂
spike whitish, filiform; tip of stamen dark,
globose; pollen grains in tetrads   .   .   .   3. *T. latifolia*

1. **T. domingensis** *Pers.*, Syn. Pl. 2 : 532 (1807); Kronfeld in Verh. Zool.-Bot.
Gesell. Wien 39 : 163, t. 4/8 & 5/5 (1889); Graebner in E.P. IV. 8 : 14 (1900);
Crespo & Perez-Moreau in Darwiniana 14 : 419 (1967); Briggs & Johnson in
Contrib. N.S. Wales Nat. Herb. 4 : 62 (1968). Type: Dominican Republic,
without record of collector (? L, holo.)

Stems 1·6–5 m. high. Leaf-sheaths with sloping scarious-margined shoulders,
rarely rounded, median leaves purple-spotted within and up the base of the
blade; blade linear, up to 1·5 m. long, 8–13 mm. wide, with an obtuse tip and
narrow base, flat above and convex beneath in dried material, glaucescent or
green above, green beneath. Inflorescence interrupted by a 1–3 cm. long inter-
node, the ♀ spike sometimes interrupted or constricted. Male spike 17–34(–40)
cm. long, 0·9–1·5 cm. wide; bracteoles flattened, forked or laciniate above,
rarely linear, red-brown, ± as long as the stamens; stamens with white filaments; anthers 2·5–3·5 mm. long, with the connective produced into a dark
globose tip usually broader than the anther, paler when immature; mature
pollen grains free, deep primrose-yellow. Female spike 18–25(–34) cm. long,
1·4–2(–2·2) cm. wide, bright chestnut- or reddish-brown at maturity;
" pedicels " numerous, subpyramidal with distinct steps 0·5–13 per sq. mm.;
bracteoles numerous; stigma linear, scarcely broader than the style but
darker, much longer than the rest of the flower; carpodia numerous, irregu-
larly distributed throughout the spike, appearing as light patches where
massed together. Fig. 1/1–8.

UGANDA. Toro District: Semliki Plain, 23 Nov. 1935, *A. S. Thomas* 1545!; Kigezi
District: Mpalo, July 1939, *Purseglove* 896!; Mengo District: Nabugulo Forest near
Bajo, Dec. 1916, *Dummer* 3029!
KENYA. Kisumu-Londiani District: Malaget Sawmill road, S. of Londiani–Eldoret
road, 5 Dec. 1952, *Dyson* 384!; Teita District: Tsavo National Park East, Galana R.
near Sala Gate, 17 Jan. 1967, *Greenway & Kanuri* 13050!; near Mombasa, Feb. 1876,
Hildebrandt 1229b!
TANGANYIKA. Musoma District: Seronera R., 12 May 1961, *Greenway* 10178!; Mpanda
District: Mahali Mts., Silambula [Selambula], 18 Sept. 1958, *Newbould & Jefford*
2438!; Dodoma District: Manyoni, 22 Aug. 1931, *B. D. Burtt* 3392!
ZANZIBAR. Makongwe I. near Pemba, 13 Feb. 1929, *Greenway* 1418!
DISTR. U2, 4; K2–7; T1–6; P; pantropical
HAB. Swamps, dams, lakes and rivers; at high altitudes in Kenya (above 1500 m.)
usually in association with *T. latifolia*; 0–2250 m.

SYN.   *T. australis* Schumach., Beskr. Guin. Pl.: 401 (1827); N.E. Br. in
F.T.A. 8 : 135 (1901); Chiov., Fl. Somala 1 : 323 (1929); Täckholm & Drar,
Fl. Egypt 1 : 87 (1941); J. G. Adam in Bull. I.F.A.N. 23 : 400 (1961); F.W.T.A.,
ed. 2, 3 : 131 (1968). Type: Ghana, *Thonning* (C, holo., K, microfiche!)

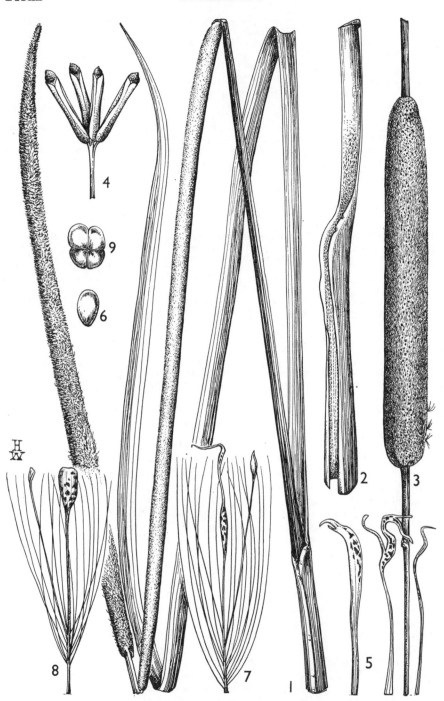

FIG. 1. *TYPHA DOMINGENSIS*—**1,** part of shoot showing mature male and immature female inflorescences, × ½; **2,** leaf-sheath, × ½; **3,** mature female spike, × ½; **4,** male floret, × 7½; **5,** male bracteoles, × 13; **6,** pollen, × 100; **7,** female floret with ovary and bracteole, × 7½; **8,** sterile female floret with carpodium and bracteole, × 7½. *T. LATIFOLIA*—**9,** pollen, × 100. 1–2, 4–8, from *Greenway & Kanuri* 12567; 3, from *Tanner* 1135; 9, from *Kerfoot* 2984.

T. angustata Bory & Chaub., Exped. Sci. de Morée 3(2): 338 (1833); Rohrb.
   in Verh. Bot. Ver. Brandenb. 11: 87 (1870); Kronfeld in Verh. Zool.-Bot.
   Gesell. Wien 39: 159, t. 4/6 & 5/1 (1889); Dur. & Schinz, Consp. Fl. Afr. 5:
   470 (1895); Graebner in E.P. IV. 8: 14 (1900); N.E. Br. in F.T.A. 8: 134 (1901).
   Types: several specimens from Greece, Peloponnisos [Morea] (P, syn.)
T. aequalis Schnizl., Die Typhaceae: 25 (1845). Type: Arabia, Wadi Hebran,
   Schimper (K, P, iso.!)
[*T. angustifolia* sensu A. Rich., Tent. Fl. Abyss. 2: 350 (1851); Dur. & Schinz,
   Consp. Fl. Afr. 5: 470 (1895); P.O.A. C: 93 (1895); N.E. Br. in F.T.A. 8: 135
   (1901); F.D.O.-A. 1: 108 (1929); F.P.N.A. 3: 12 (1955), *non* L.]
T. angustifolia L. var. *domingensis* (Pers.) Griseb., Fl. Brit. W. Ind. Is.: 512
   (1864) & Cat. Pl. Cub.: 220 (1866)
T. angustifolia L. var. *australis* (Schumach.) Rohrb. in Verh. Bot. Ver.
   Brandenb. 11: 83 (1870)
T. angustifolia L. subsp. *domingensis* (Pers.) Rohrb. in Verh. Bot. Ver. Brandenb.
   11: 97 (1870)
T. angustata Bory & Chaub. var. *leptocarpa* Rohrb. in Verh. Bot. Ver. Brandenb.
   11: 88 (1870); Kronfeld in Verh. Zool.-Bot. Gesell. Wien 39: 161 (1889);
   Graebner in E.P. IV. 8: 14 (1900). Type: Ethiopia, Tigre, Djeladscheranne,
   Schimper 1563 (K, iso.!)
T. angustata Bory & Chaub. var. *aethiopica* Rohrb. in Verh. Bot. Ver. Brandenb.
   11: 89 (1870). Type: Ethiopia, Takazze, *Schimper* 1190 (P, holo.)
T. angustata Bory & Chaub. subsp. *aethiopica* (Rohrb.) Kronfeld in Verh. Zool.-
   Bot. Gesell. Wien 39: 162 (1889)
T. angustifolia L. subsp. *australis* (Schumach.) Kronfeld in Verh.
   Zool.-Bot. Gesell. Wien 39: 156, t. 5/4 (1889); Graebner in E.P. IV. 8: 13 (1900)
T. angustata Bory & Chaub. var. *abyssinica* Graebner in E.P. IV. 8: 14 (1900),
   nom. illegit. Type: as *T. angustata* var. *aethiopica*
T. domingensis Pers. var. *australis* (Schumach.) J. B. Gèze in Bull. Soc. Bot. Fr.
   58: 459 (1911) & Etudes sur les Typha: 118 (1912)
[*T. latifolia* sensu F.P.U.: 209 (1962), *non* L.]

NOTE. The above synonymy does not cover the whole range of this species, more
detailed accounts can be found in Gèze, Etudes sur les Typha (1912), Adam in Bull.
I.F.A.N. 23 (1961) and Briggs & Johnson, Contrib. N.S. Wales Nat. Herb. 4 (1968).

2. **T. capensis** (*Rohrb.*) *N.E. Br.* in Fl. Cap. 7: 32 (1897); Hiern, Cat. Afr. Pl.
Welw. 2: 85 (1899); N.E. Br. in F.T.A. 8: 136 (1901); F.D.O.-A. 1: 108
(1929). Type: South Africa, Cape Province, Uitenhage, *Ecklon & Zeyher* 913
(SAM, lecto.)

Stems erect, stout, 2–4 m. high. Leaf-sheaths with abruptly rounded or
auriculate scarious-margined shoulders, sometimes purple-spotted within;
blade linear, up to 2 m. long, 10–15(–20) mm. wide, with an obtuse tip and
narrow base, flat above and convex on the back in dried material, glabrous,
glaucous. Inflorescences contiguous in tropical African material, rarely
separated, each subtended by a caducous foliaceous bract, rarely the ♀ spike
constricted or interrupted and each part with a bract and separated by up to
2 cm. Male spike 8–11(–15) cm. long, 1–1·2 cm. wide; bracteoles flattened
and forked or laciniate above, rarely linear, red-brown, ± as long as the
stamens; stamens with white filaments fused at the base; anthers ± 3 mm.
long, with the connective produced into an obtuse, subtriangular dark tip as
broad as or narrower than the anther, pale when immature; mature pollen
grains free, deep primrose-yellow. Female spike (12–)14–32 cm. long,
1·4–1·8(–2·3) cm. wide, bright chestnut- or reddish-brown at maturity,
spotted with off-white if carpodia are abundant; "pedicels" numerous,
filiform; bracteoles occasionally present, never numerous; carpodia pale,
speckled with red, ± as long as the perigonous hairs; fertile gynoecia with a
narrowly lanceolate stigma much exceeding the perigonous hairs.

UGANDA. Kigezi District: Kachwekano Farm, June 1951, *Purseglove* 3641!; Mengo
   District: Kyagwe [Kiagwe], Namanve Swamp, Apr. 1932, *Eggeling* 247 in *F.D.* 659! &
   Kampala, King's Lake, Mar. 1936, *Hancock* 175!
TANGANYIKA. Maswa District: Lake Victoria, 6 Mar. 1965, *Leippert* 5612!; E. Usambara
   Mts., 1·6 km. Amani–Monga, 26 Dec. 1956, *Verdcourt* 1727!; Uzaramo District:

Dar es Salaam, Feb. 1874, *Hildebrandt* 1229 ! ; Songea District : Mshangano fish-ponds, 21 Mar. 1956, *Milne-Redhead & Taylor* 9311 !

Zanzibar. Zanzibar I., Kama, 17 Apr. 1962, *Faulkner* 3034 ! ; Pemba I., Chake Chake, 7 Oct. 1929, *Vaughan* 691 !

Distr. U2, 4 ; T1, 3, 6, 8 ; Z ; P ; Congo and throughout southern Africa

Hab. Swamps, dams and rivers ; 0–1950 m.

Syn.   *T. latifolia* L. subsp. *capensis* Rohrb. in Verh. Bot. Ver. Brandenb. 11 : 97 (1870) ;
Kronfeld in Verh. Zool.-Bot. Gesell. Wien 39 : 180, t. 5/13 (1889) ; Graebner in E.P. IV. 8 : 10 (1900) ; Anderson in F.S.A. 1 : 54 (1966)

[*T. latifolia* sensu Engl., P.O.A. C : 93 (1895) ; N.E. Br. in F.T.A. 8 : 136 (1901), pro parte ; F.D.O.-A. 1 : 108 (1929), *non* L.]

3. T. latifolia *L.*, Sp. Pl. : 971 (1753) ; Rohrb. in Verh. Bot. Ver. Brandenb. 11 : 75 (1870) ; Kronfeld in Verh. Zool.-Bot. Gesell. Wien 39 : 176, t. 5/11 (1889) ; Dur. & Schinz, Consp. Fl. Afr. 5 : 470 (1895) ; Graebner in E.P. IV 8 : 8 (1900) ; N.E. Br. in F.T.A. 8 : 136 (1901), pro parte ; Chiov., Racc. Bot. Consol. Kenya : 123 (1935) ; Crespo & Perez-Moreau in Darwiniana 14 : 416 (1967) ; F.W.T.A., ed. 2, 3 : 131 (1968). Type : from Sweden

Stems erect, stout 1·5–3·5 m. high. Leaf-sheaths abruptly rounded to auriculate at the junction with the blade and scarious-margined, sometimes purple-spotted within ; blade linear, up to 2 m. long, 8–10 mm. wide, with an obtuse tip and narrow base, the base flat above and convex beneath in dried material, glabrous, glaucous. Inflorescence of contiguous spikes, each subtended by a caducous foliaceous bract, the ♀ spike sometimes constricted or interrupted, each part with a subtending bract and separated by 0–3·5 cm. Male spike 8–12 cm. long, 1·3–1·8(–2·5) cm. wide at maturity ; bracteoles linear, beige or pinkish-fawn, ± as long as the stamens ; stamens with white filaments stouter than the bracteoles ; anthers 3–4 mm. long with the connective produced into an obtuse rounded black tip as broad as the anther, or broader ; mature pollen grains adhering in tetrads, pale primrose-yellow. Female spike 13–16(–25) cm. long, 2·5–3·5(–4) cm. wide, yellow-green at first, becoming dark sepia-brown or almost black at maturity ; pedicels numerous, filiform, 20–30 per sq. mm. ; bracteoles not present ; carpodia pale, speckled with red, ± as long as the perigonous hairs ; fertile flowers with a broadly lanceolate stigma much longer than the perigonous hairs. Fig. 1/9, p. 3.

Uganda. Elgon, Budadiri, 8 Dec. 1927, *Snowden* 1223 !

Kenya. Nairobi, Karen Estate, Mbagathi R., 14 Jan. 1934, *Napier* in C.M. 5823 ! ; Kisumu-Londiani District : 2·5 km. on old Londiani–Eldoret road, Dec. 1952, *Dyson* 385 ! ; Kericho District : Sotik, July 1962, *Napper* 1680 !

Distr. U3 ; K3–5 ; throughout the north temperate regions of the world, southwards to Kenya and Nigeria in Africa and to the Argentine in America

Hab. In swamps, dams and rivers with permanent flow ; 1300–2200 m.

INDEX TO TYPHACEAE